Evolution AT ITS BEST

MARTIN STUBBS

PAGE PUBLISHING
Conneaut Lake, PA

First originally published by Page Publishing 2024

ISBN 979-8-88793-946-9 (pbk)
ISBN 979-8-88793-990-2 (digital)

Printed in the United States of America

CHAPTER 1

Evolution

Evolutions are inheritable traits that are in a population that spreads across generations. Traces of evidence, fossils (bones of the past), were found to prove evolution exists in Africa, some parts of Europe, and even in New Jersey, USA. It has been proven that evolution has derived from something and becomes something new.

Scientists have proven that the structure has to be there and then evolves into something new. Evolution is a step-by-step process that causes changes in one's life, whether it is conceptually or experientially.

Since over two billion years ago, 1.2 million species have been documented, and 86 percent have not been described. In May 2016, one trillion species were on earth, and only one thousand of 1 percent were described. Evolution is one of the wonders of nature that trying to keep up is one of the most difficult tasks scientists had to face. The change from DNA to proteins, body structure, and characteristics is certainly remarkable. It is amazing that tracking so many life-forms, some small in nature, is very difficult.

Adaptation is getting used to our surroundings and biology. The universe is always one step ahead of what we need for that very moment.

An evolving body structure gives flexibility; evolving brains can be useful for survival. We are here to dominate and be masters on earth.

Evolution over two billion years has given us various climate changes and volcanic eruptions that create a system of life into new life.

Natural selection is a correlation to adaptation, which was first highlighted by Charles Darwin, and is the master indicator of organisms adapting to biology, surrounding and producing offspring.

Charles Darwin went on a voyage for five years, beginning in 1831–1836, and picked up fossils and plants along the way to prove his theory.

He showed the world how evolution is a true concept and that it correlates to natural selection. He proved with written theories of evolution that life-forms came from a single-cell organism. Other scientists and naturalists up to this day have respected his work and truly believe in his theories. Scientists are not sure how long-ago life started, but they know that evolution took a course of over two billion years.

Geologists

I cannot go on with evolution without mentioning the geologists who play a major part of evolution. Geology is the study of rocks to determine and record the times and things that occurred in the past.

First is **Nicolas Steno**, **Danish**, who discovered four rules of geology:

1. The law of original horizontality means the rocks that are horizontal have to be placed flat before they become horizontal.
2. The law of superposition means that the rocks to the bottom are older than any previous surface.
3. The law of crosscutting means that any layer of rocks that is tipped had to be flat before it was tipped.

4. The law of faunal succession means any layer of rock that has plants, vegetations, or fossils is older than the bottom surface.

He was honored by Queen Victoria on his discovery of dating rules.

Second is **1700's Giovanni Arduino** on nepotism naming layers of rocks determined by what surface you are standing on, which becomes the most recent in time. The farther you go below the earth, the older it becomes.

For example, if there are four layers, the layers are determined as follows:

- The quaternary layer is the surface you are standing on.
- Tertiary is the second surface layer from the top.
- Secondary is the third surface layer from the top.
- Primary is the fourth surface layer from the top.

In **1739–1839, William Smith**, a **London** surveyor of property lines, discovered how to determine time of fossils by drawing geologic maps.

In **1726–1797, James Hutton** proposed plutonism, which described that rocks were formed by water.

In the 1700s, **George Cuvier** proposed **catastrophism** on disasters that has a formality that never changes.

Charles Lyell is a great writer on uniformitarianism. His theory was that nature has a way of doing things where it remains the same with no changes, which they called **uniformitarianism.**

In 1750–1817, **Abraham Werner** named the layers of rocks primitive, transitional, secondary, and tertiary. Abraham says all rocks were formed from water.

Evolution is a teacher to the purest form of life. It had a slow development from over 4.5 billion years ago. Most scientists, geologists, and naturalists have confirmed that life started around that time. The times of extinction as covered by scientists show that there were signs of uniformitarianism.

That is true for the years from the time fish started to move from sea to land. This will be explained in detail in the next chapter, which is the findings of geologists Nicolas Steno, Giovanni Arduino, and William Smith. Nicolas Steno was called the father of stratigraphy.

Stratigraphy is the study of earth's history recorded in rocks.

Nicolas Steno is the trendsetter of geology: he set the laws, and geologists follow them.

There were theories that were argumentative; one of those theories was catastrophism, which described that catastrophes happen at certain times, causing extinction. This was the theory made by George Cuvier, and the other theory was uniformitarianism, which explains that there are certain times when uniformitarianism takes place, and there are no changes to the time which causes extinction as discovered by Charles Lyell; his theory won that argument.

The other theory was where rocks came from; Abraham Werner, a German geologist, says that rocks came from water.

Rocks really come from a cool liquid known as lava from volcanic eruptions and have two kinds: **the intrusive rocks**, which are rocks that cool on the inside of the volcano, and **the extrusive rocks**, which are cool molten rocks outside volcanoes. Lava cools and develops a hard crust to form igneous rocks. **Igneous rock** is just cool molten rock.

Sedimentary rocks are found near rivers and streams mixed with minerals and form layers caused by heat and contact pressure. There are three different kinds of sedimentary rocks: **Clastic rocks** are rocks classified according to sediment size. **Crystalline rocks** are formed from evaporation or precipitation of minerals in water. **Bioplastic rocks** are formed from the compaction and cementation of organic matter.

Metamorphic rocks are rocks formed by contact and heat with new formations.

Metamorphic rocks are the only rock that has bands of light and dark. The bands are formed under intense pressure minerals lined up. Minerals are inorganic materials that are mixed in from other rocks.

The recycling of rocks is amazing. This nonending process starts with the pushing of rocks and sand below the earth.

The intense pressure causes a liquid to form, called lava or magma. The heat gets so intense with pressure that it explodes out from the mantle or volcano and forms igneous rocks. Sedimentary layers form through the travel of streams and rivers. Metamorphic rocks are formed from different minerals in the water. The next chapter explains existence in more detail and how life-forms came about.

CHAPTER 2

Existence

Existence is the fact or state of living or having objective reality.

I am about to give an incredible story of how scientists have an interesting theory of life existence, according to scientists.

It is clear and credible today. Scientists are not sure how and when life began but develop theories that have been proven. With over four billion years of twists and turns between climate changes, life forms where extinctions became mundane. It all started over a course of two billion years ago when the earth was bombarded by asteroids and comets. Below the earth's surface, there are gases trapped, causing molten rock to be formed.

These gases that came from asteroids and comets hitting the earth not just changed the geographical plain but also created a phenomenon that gives an origin to all life-forms. All origins of life-forms came from a single-cell organism.

A combination of heated rock and ash in earth's space caused comets to form water here on earth. A lightning blast with over one billion volts of electricity hit the water at the right time, forming atoms with a chain of genetic material to form one-cell organisms. The cells had to face climate challenges, so the water chemical and compounds form an oily covering over the cells.

The cells started to merge, sending out messages of genetic material to other cells that made copies of themselves. Scientists say that there was a mistake when two cells merge to form a single cell, which created male and female genetic material. This was when sex was born.

Mutations of these cells when they copied themselves cause variations of species.

Every single life-form could be traced to that one single-cell organism. The cells continue to mutate until our ancestors became a three-foot water worm. There was a problem because they were all blind, and finding a mate was hard. Over a course of two million years of evolution, a handful of skin cells mutated to form eyes. This led to light-sensitive skin cells to continue to mutate and kept refining and developing the form of eyes. Evolution has proven that it is a step-by-step process. Eyes were developed gradually because of mutations. Mutations are the changing of genes to alter, modify, and create variations. Our ancestors became fishlike creatures called **Milo Canengia**. Then we had to make sense of what we see, so behind our eyes, nerve cells mutate no bigger than a pinhead to form the brain. Milo with a brain has new senses. Even with a brain, we could not outwit our predators. Our ancestors became prey for the great whites in the Indian Ocean.

Over two million years of evolution, we became a **one-foot armored fish** with large teeth and a jaw with pure eating power. We were chased by great whites of the Indian Ocean to shallow waters, where some of our ancestors died of carbon dioxide. With saturation and lack of oxygen, they could not swim back into the deep ocean. A new lifeline was given to us. It is called natural selection, which helped us to adapt to our surroundings and biology. That is when we started to breathe air four million years ago.

Our next step is to survive in water and land. That was when a new organ was formed: **gills** were used to shut down our windpipes to prevent us from taking in too much water when we are in water. We became *Ichthyostega*, a fishlike creature that survives on land and water. We had an evolving brain, larger teeth, skin to face the sun, feet to help raise our lizard-like bodies, and toenails, which then developed to be our fingernails to face tough surfaces. We developed tougher skin to face the sun, and we became **Carceneria**, the first land creature. These developments happen over a period of millions of years. The next step is to increase the species. The male cannot lay eggs, so the offspring cannot be deposited into a female before the egg is formed, but the male can deposit the offspring before the egg is formed in the female, and the offspring will be fed through the term by the embryo. This is sex as we know it.

The evolution of other species was developing all at once. At this stage, there are more than one million species developed. The new offspring of Carceneria had evolving brains, where they adapt to new sights and scents and the brain has more to process, and Carceneria became the first land animal.

Carceneria evolved into a lizard-like creature called **Berenox** with large jaws, bigger teeth, and new body structure. Berenox became a predator, and its prey was a new species that evolved from a reptile called *Protosaurus*.

When we thought the earth was finished, with extinction far, far away, magma was forming below the surface of Siberia. Pulling it apart, with space as big as the United States, one thousand feet deep below, surface volcanic eruptions happen for over thirty million years. Most of the species died except us, and a few species survived. Our ancestors became **Ectominione**, a cat-sized creature with fur. A new species had evolved from reptiles that survived the volcanic eruptions, and it brought us the first dinosaur called *Herrerasaurus*. *Herrerasaurus* became the dominant force in the Montana forest; it developed so quickly and adapted to the new world rapidly. The evolution process continues as the Ectominione got smaller over millions of years to avoid being eaten by dinosaurs, and with a sharper brain that can see and hear a dinosaur before a dinosaur sees or hears them, it becomes a shrewlike creature called **Batadone**. The Batadone was evolving by increasing its brain capacity to have sharp senses to avoid being eaten by dinosaurs. Batadone was the first to bear live offspring and nurture them by feeding their offspring milk.

Then our ancestors were faced again with another disaster ten thousand miles off the Yucatán Peninsula in Mexico when an asteroid hit, causing 95 percent death of species that once lived.

The only species that survived is us and a few others. What was a local disaster became a global disaster. Volcanic eruptions happened all over because of hot molten rock beneath the surface.

Most affected were the dinosaurs because they are big creatures; most of the plant eaters and carnivorous eaters died. Dinosaurs that reigned for 165 million years now had nothing to eat because all vegetation dried out, and the temperatures were over one hundred degrees; the sun was blocked out by ash. The dinosaurs couldn't adapt and died.

After the disaster, bugs surfaced to the top surviving on decayed bodies. Our ancestors were still Batadone, a ratlike creature. Once before was another species of mammals called shrew but is now a rodent. We became a bug-eating rodent called **Purgataurus.**

Fifty-six million years ago, trees were formed with ripened fruit. From having life on the ground, we started life on the trees. Natural selection intervenes as an assist to adapt us to our new world, causing our tails to be concealed in our spines, from being Purgataurus to a treelike creature called **Altio Gracias**, new species of mammals called primates. Our ancestors' new life is on trees, leaping from tree to tree to get access to fruits filled with nutrients; the more fruit we eat, the longer we live. Mammals became the dominant species, spreading to each continent.

Over ten million years ago, there was a disaster again. The African and Arabian plates in Siberia below the earth were pulling apart, causing the formation of the African rift valley to rise 3.5 thousand miles up, blocking the Indian Ocean from watering the lands of Africa. The forest was ravaged and dense; trees were far apart and hard to reach. At this time in East Africa, we became **Ramapithecus Ramadis**, apelike creatures. Ramapithecus can walk on branches and move along on branches with their hands and have a brain as big as a grapefruit.

The trees were far apart, so Ramapithecus had to learn to let go. Two million years ago, Ramapithecus took his first steps on two legs to go to another tree and use his hands to pick fruit. It was easily picked up by Ramapithecus, and it was passed down to their children and others. Over millions of years of evolution, we then became

Australopithecus, with a brain the size of a grapefruit. Over two million years ago, we walked on two feet. We evolved to walk faster and had no problem finding a mate and food.

Our ancestors were faced with birth problems. Babies have to come out early as the pelvis is small, and the heads of babies have to be small to pass through the pelvis. This is why we take care of our young, protecting and feeding them, because two million years ago, we had premature babies. Evolution continues the rising of mankind. After 2.3 million years, the first man arrived, ***Homo habilis***. This species of mankind had sharper brains and are intuitive, and they were called handymen.

The skills of *Homo habilis* were extraordinary; he made tools to protect against predators and hunt for food. These early stages of evolution of mankind all occurred in Africa.

Our family tree was lost, and scientists couldn't find enough fossils (bones just below a billion years old) to continue to piece their story together. The loss of bones became a problem for a little while. Finally, they found a missing piece to the puzzle, ***Homo erectus***.

Homo erectus came from Africa and had superintelligence. They worked together and were great hunters. They have the body of a new species that is built for hunting; going against faster prey was a show of their excellence: they can run. Low shoulders and long torso stabilize them; they have buttock muscles that can expand and push them forward and also sweat glands to prevent overheating. Their prey may be faster, but *Homo erectus* could run farther. Their prey gets tired because fur develops heat, and they fall with exhaustion.

One day over 1.2 million years ago, *Homo erectus* found fire after lightning struck grasslands close to where they were hunting. It struck the idea that fire can assist them with supplying warmth, light, and protection and cooking raw meat. *Homo erectus* got around to cooking when they had an evolved brain, the final piece to the evolution of *Homo erectus*, the most cunning hunter there was in his time. The *Homo erectus* family would sit around together, eat together, and get warmth from fire altogether. *Homo erectus* was the first human species to be united; they did everything together.

One more intervention of natural selection happened. Our tongue changed to be flexible, with the larynx down our throats; the tongue changed shape to make sounds and finally words. We finally reached the point of supremacy when we became ***Homo sapiens***, meaning "wise men," who also came from Africa. We are a young species of 220,000 years old with six thousand years of civilization.

We are superintelligent and are the masters of the world. The next chapter gives you more insight into other human species that existed millions of years ago, but the characters mentioned in this chapter are an indication of mankind rising.

CHAPTER 3

Other Human Species

The human species had a total of fourteen species of *Homo sapiens*: there are six descriptive species known by name with no real history, three documented species that are fully described, and five species lack history. The reason is that the species that were given names and lacked history did not have enough fossils (past bones) to record time, social systems, and survival. I would begin by stating the three species that are fully described, starting with the **Neanderthals**, who lived for three hundred thousand years and went extinct. The males were five feet two, and the females were five feet one. The theory from scientists is that not much fossils (past bones) were left for investigation. The little they gave us still can paint a decent picture of their lifestyle. They did find fossils in the Atlantic Ocean and Western Europe. Scientists discovered that they migrated a lot throughout their existence.

They would continue traveling in groups between Europe, France, and England, even Asia. One of their main settlements was on an island named Jersey located near England's landscape. Prolanthrogists, who form the faces of prehistoric people, discovered that Neanderthals had big brains, noses, and lungs.

Their Neanderthal males were five feet two, and their females were five feet one. Their nose served as a humidifier. It brings in a lot of oxygen, which is evenly distributed to assist with energy.

Neanderthals use a lot of energy because they hunt large wild animals like bisons, mammoths, and rhinos. They would use their prey for food and their skins for clothing. They would plan attacks by catching these wild creatures at rivers and streams where they drank water and the surface was soft. This gave them an edge because it would give their prey a very slow takeoff if they had to run. Neanderthals were there ready with weapons to attack.

Their style of hunting was very cerebral; they would have a place to kill their prey and another to store the carcasses.

Homo sapiens were interacting with Neanderthals over 175,000 years ago because it was documented that they found *Homo sapiens* genes in Neanderthal DNA. Neanderthals adapted to their world of low temperatures below -20 degrees Celsius. Scientists also discovered that they had meetings with *Homo sapiens* to trade females on both sides to enhance genetic continuity. Their advanced weapons were also traded, and from archaeologists' findings at the time, *Homo sapiens* met with the Neanderthals just when they became extinct. After that, *Homo sapiens* were widespread all over. Scientists are not sure how they became extinct, but one of their theories mentions that they may have territorial war. *Homo sapiens* outnumbered them and took over. Neanderthals became extinct forty-two thousand years ago.

The next human species that they have information, which I can share, is ***Homo floresiensis***, which was discovered in 2003 on the Indonesian island of Flores. Fossils found show that they had tiny brains and large teeth and lived about fifty thousand to one hundred thousand years ago. They had stone tools to assist with making weapons. On the island of Flores, the animals there were unique in size.

They would have very large birds, rodents, and Komodo dragons. Scientists were not sure if *Homo floresiensis* was prey.

Scientists want to believe that volcanic eruptions may have wiped out the species. After all, there are signs of volcanoes present.

Homo rudolfensis was discovered in 1986 in eastern Africa, northern Kenya, northern Tanzania, and Malawi between 1.8 and 1.9 million years ago. There was really one good fossil which was a skull of *Homo rudolfensis*. *H. rudolfensis* was compared to *H. habilis* by the braincase, but they thought that there wasn't a fair assessment that it can be the same because of the size of the molars. It was compared to the species *Australopithecus* because of its large brain.

The name *Homo rudolfensis* was given by a Russian scientist called V. P. Alexeev in 1986 after Richard Leakey's team. Richard Leakey's team named him after a lake called Rudolf, which was called Lake Turkana in 1972.

Homo heidelbergensis was discovered in 1908 in Europe, Asia, China, and Africa. They lived two hundred thousand to seven hundred thousand years ago.

For their height, males were five feet nine and females were five feet two; for weight, males were 125 cm, and females were 157 cm. They use stone hand axes as tools. Other tools include Katanda bone, wooden spears, punctured horse shoulder blades, stone sickle blades, and projectile points.

Australopithecus was around 1.2 million years ago. They were a member of the human species that walked upright all the time. The name is synonymous with names like ***Paranthropus*** and ***Kenyanthropus***. The species lived in East Africa and had a huge brain. Eventually, they became extinct.

Ramapithecus was around 5.3 to 16.6 million years ago. They were around East Africa on trees, but because trees were far apart, Ramapithecus had to take its first step on two feet to go from tree to tree and use its hands to pick fruits. Ramapithecus taught their children to walk and pick fruit. Ramapithecus evolved to become *Australopithecus*.

Denisovans started when they found fossils of a young girl at Denisova Cave in Siberia in the Altai Mountains, Russia, and realized when tested that she had genes of Neanderthals. They found fossils of a male molar which happened to be a Neanderthal and fossils of an adult female and realized she was a Denisovan. Scientists put it together that between 2003 and 2010, they had found fossils of a family. This is how they got the name Denisovans from the cave Denisova, where they were found. Denisovans used to interbreed with Neanderthals (a human species). Scientists found out that there were a lot of meetings between these species of humans. The girl they found first, which is the daughter, lived from 52,000 to 76,000 years, which makes her twenty-four years old. It wasn't discovered how old her parents were.

We had access to information about these species through their fossils given to scientists, searching and looking for fossils of our species to this day.

It is difficult because, without fossils, information cannot be given out about species. I discovered after following evolution for over a billion years that, no doubt, there was some sort of formality to this process. The universe always throws us a lifeline of natural selection to help us to adapt to our environments, whether it is body structure or intuition to survive.

At this time, fossils are needed to gather information. Archaeologists are still in Africa and Denisovan Cave in Siberia looking for fossils to either confirm or make something new of their findings. The history of mankind is really an ongoing investigation because of missing pieces to the puzzle of mankind. For instance, I just found out that the Denivisovans rode a certain breed of zebras over thirty thousand years ago.

Mankind was judged through certain criteria to assess progression, for example, cranial size, random walk, gradualism, stasis, punctuated equilibrium, stasis-random walk, and stasis-gradualism. These are categories on which scientists based their theories of progression or nonprogression. Then their findings would branch to locomotion, cognition, dexterity, and diet. We know that there were a lot of interactions with Neanderthals, and because of *Homo heidelbergensis*, genetic material was found.

The universe always wanted to take care of all its living things, but those species that went extinct really shape who we are and get us to where we are today. We've become the undisputed masters of the world.

CHAPTER 4

The Reigning of Dinosaurs

Dinosaurs reigned on land, sea, and air for 135 million years. Dinosaurs evolved for millions of years into different species of reptiles. Firstly, dinosaurs evolved from a reptile called **Archosaurs**. Archosaurs are then broken into two groups: *Saurischia* and *Ornithischia*.

Saurischia are dinosaurs that have their pubis bone facing forward and down (e.g., tyrannosaurs). *Ornithischia* are dinosaurs with their pubis bone facing backward and down (e.g., hadrosaurs and megasaurs).

Dinosaurs travel in herds for protection and give support as these travels were survival of the fittest. For dinosaurs, when they want a mate, it becomes a competition among other males; they would create a nest for themselves and even show off food to get mates.

The dinosaur migration trips, especially at night, were the most dangerous. The lack of food while traveling for long distances encourages predators to attack the most vulnerable dinosaurs. The most treacherous dinosaurs were the **tyrannosaurs**, 13 m long and 3.8 m high, with large teeth, meat eaters, weighing six tons, and strong predators. Every piece of vegetation found along the way was always a fight. Tyrannosaurs evolved from iguanodon.

Iguanodon means **"broken tooth,"** which was given its name by the finding of a large tooth by paleontologist Gideon Mantell in the 1800s. Mantell started the reconstruction but only had a tooth. Then many years later in the coal mines of Belgium, they found skeletons of iguanodon. They use what they found to begin reconstruction to give a clear idea of how this creature looked. Iguanodon had no front teeth, had a lot of two-inch back teeth, and was a plant eater thirty feet long and sixteen feet high. They had battles with megasaurs.

Megasaurs were land dinosaurs; they joined the herds of dinosaurs when they migrated many years later and evolved to the sea and became great predators of the sea and prey for the great whites. Migration was important to lay eggs.

Scientists well documented that the greatest and smartest dinosaur was the Troodon. **Troodons** was discovered in 1993 by paleontologist **Crate Dersler**, who gave the name **dragon's grave**. It is a fossil-like grave in East Wyoming known for finding dinosaur fossils for over a century. Troodons were dinosaurs with a brain the size of humans'. They had large eyes facing forward, which indicate that they were great hunters for the night, and had special claws for holding their prey.

Paleontologist Dale Russell from North Carolina State University has been hunting for fossils for decades right there at the dragon's grave. He found fossils of duck-billed and Troodon dinosaurs at this very spot in East Wyoming. Another place for hunting dinosaur fossils is in New Mexico. It just seems that nature continues to throw lifelines for dinosaurs until the very end. I believe that it was their time for extinction after reigning for 165 million years. Dinosaurs were remarkable creatures; they all had their time to fight among each other. There were some brutal battles shown through scientific investigations. Nature, as I see it, is sacred to free will because all this occurs from a single-cell organism thrown at us by natural selection. It helps all living things to adapt to their surroundings and biology. I cannot forget to list the ten most ferocious sea dinosaurs that rule the sea, starting from position 10 to 1:

10. ***Plesiosaurus*** lived in Norway 150 million years ago; had a massive head, broad flippers, and short necks; were forty feet long; weighed thirty tons; and had a strong bite as a *Tyrannosaurus.*

9. ***Kronosaurus*** evolved from *Plesiosaurus*; fossils were found in Australia and Columbia. It had a large head and jaws, thick squarish trunk, and thirty-four-foot-long snap to tail. It was a fierce predator weighing twelve tons. This dinosaur could go at high speeds to catch its prey and have a super bite.

8. ***Nothosaurus*** evolved from Plesiosaurs, which lived 230 million years ago. They were fifteen feet long, have long needlepoint teeth, and are not proven to lay eggs. They are also called false lizards, and they came on land to rest and eat squids and small fish.

7. ***Stegosaurus*** evolved from Plesiosaurs and lived seventy to eighty-five million years ago with a snake-like body thirty-five feet long and weighing four tons. They have long necks and cone-shaped teeth.

6. ***Albertonectes*** lived seventy-six million years ago, evolved from *Plesiosaurus*, had small heads, and was thirty-eight feet long with a neck covering twenty-three feet. Its neck had seventy-six bones; it ate squids and fish. In order to digest its food, the stomach had stones, which are also called gastro lids, to crush the food by rubbing its stomach on the sea's floor.

5. ***Thalassomedon*** was also called sea lord and evolved from *Plesiosaurus*. It was forty feet long, its head was nineteen inches long, and it had long flippers and fifty-two-inch sharp teeth.

4. ***Tylosaurus*** is family to the mosasaurs, which started off as a land dinosaur that evolved to become a sea dinosaur, which lived eighty million years ago and had a narrow but powerful head, long tail, and agile flippers. This dinosaur was a carnivore and swam in shallow water. Its prey were flightless birds, turtles, mosasaurs, plesiosaurs, and small sharks.

3. ***Shonisaurus*** lived 215 million years ago; fossils were found in 1920 in Nevada. It had dolphin-like appearance, was fifty feet long, weighed thirty tons, and had no teeth. It is documented that its young had teeth but fell all out while maturing.

2. ***Mosasaurus*** lived sixty-six million years ago; it had a crocodile-like head, was fifty feet long, and had teeth on both sides of the jaw. Prey was fish and squid.

1. ***Shastasaurus*** lived 250 million years ago, was sixty-nine feet long, weighed seventy-five tons, and was heavy and tall as a blue whale. If it stood up, it was as high as a seventy-story building. Prey were octopuses, squid, and fish.

I think species should be defined as a group of entities that came from the same origin. Dinosaurs have one thousand species, and the thought of that is mind-boggling. If you were to think of the kind of ecosystem they face, it is really hard to take, including rapid climate change and volcanic eruptions as the asteroid hit earth. Dinosaurs shape who we are. Supremacy really was signatured earlier on by dinosaurs. Everything that has happened on earth was of no accident; it was all by design to get *Homo sapiens* here and presently calling the shots.

Purpose plays an important part in nature, looking at how the grounds were laid out for vegetation. Providing food for the dinosaurs, that system was a foundation for life-forms to multiply itself.

Coincidence is fiction, but purpose is real.

Evolution is such a wonder of the world; nature provides enough just to fill the void needed. We are all fortunate to be given life.

Natural selection helps us to adapt to our new world. Fossils are constantly found all over the world.

There is so much missing in discovering life-forms. Scientists are always busy either in caves all around the world near rivers and streams or on sedimentary rocks. What we think that we are on a platform that has no depths is not true. There are many platforms beneath us that go millions of years ago. Scientists are exploring new places around the world where they find fossils. The curiosity and obsessiveness continue with these scientists, paleontologists, and archaeologists.

The popular places to find fossils are China, Australia, Columbia, and America, especially in dragon's grave, Wyoming, USA, and the Denisova Cave in Russia. Natural selection plays so much of a role in evolution. I would explain to you what natural selection is to give you a clear picture.

Natural selection is the adaptation of surroundings and biology.

The reason I give the definition of it is to give a clear understanding of what is coming next.

The factors that prove evolution exists are as follows:

1. **Direct observation**: which mutations pass on
2. **Homology**: natural similarities in certain species
3. **Molecular homology**: genes found in bacteria
4. **Biogeography**: the finding of certain fossils in certain areas
5. **The propagation of genetic variance**: the production of variances in genetic material

Natural selection, which was founded in the eighteenth century by Charles Darwin, was proven by the factors that are listed above. He laid the foundation for all scientists to build on to prove all theories.

Then once proven, there was no looking back; all theories were followed up to the present time. Charles Darwin became a trendsetter in everything involving evolution.

The choice of which species survive or become extinct was a blind choice of nature. Creating genetic material to either weaken or strengthen a species is remarkable and amazing at the same time. There is some degree of expansion in species and elimination of species through genetic material.

Climate change is also a way to increase species or shorten the life spans of species. Certain temperatures, for example, hot and cold temperatures, could shorten life spans and cause extinction while temperatures of the tropics can increase variances of species and lengthen the life of species.

Fossils give scientists a record of proven life and length of existence.

This all happened because of Charles Darwin, the pioneer. The mention of his laws was listed in chapter 1. Imagine it was Charles Darwin who said and proved that every living thing came from a single-cell organism.

Dinosaurs faced a lot of catastrophes even though they were fast; they never evolved with the right intelligence to match. Natural selection started to subtract some things from dinosaurs and only had the Tyrannosaurus, the only dinosaur that matches their brain to the size of a human brain.

Rapid climate change, volcanic eruptions, and change in genetic material over millions of years evolved dinosaurs into birds. This was proven by the slow change in the formation of bones and genetic material. Scientists have mathematical formulas to learn how and why dinosaurs went extinct. These formulas help to understand, but those weren't the entire answer because there is a lot to the equation of how evolution works.

Scientists also use a clock for speciation and timing to determine how nature works by balancing the species. Diversification could be used to increase species by filling niches, or it can reduce species as well by species having battles among themselves, which can cause the extinction of that particular species before they could have offspring.

Mutations cause variations of distinct characteristics and body structure.

CONCLUSION

All life-forms came from a one-cell organism. It is also amazing that the human species changed so many forms from fish to rodent and then mankind. We survived for over two billion years of evolution. Nature always seems to have our backs in surviving. All the geologists became curious to find what they did to fill some of the voids. In the forest of Montana, human species came from a small rodent. The body structure of a rodent helps us to hide in small places in the forest and avoid being prey.

Dinosaurs were there, and it made us competitors in the forest. Imagine being a small creature just moving around with natural selection, providing us with better senses to see and hear dinosaurs long before they get to us, instinctive enough to pass on genes to cause mutations that trigger variances of producing different species, having different body structures to adapt when we were on trees and picking fruits from trees to evolve into mankind having the superintelligence we have now. The idea of knowing that the human species is only 220,000 years old and that civilization only exists six thousand years is amazing. We are a young species that I think will only keep evolving with a population of 7.9 billion, and the multiplication will continue. Natural selection keeps throwing the human species lifelines to adapt because our superintelligence provides the tools to work for us.

My name is Martin Stubbs, and I was born in Trinidad, West Indies. My newly found passion for writing in the year 2020 via a vision gave me the title of the book—*Evolution at Its Best*. I then began my journey as an author residing in Newark, New Jersey. I think it was a definite sign of what my purpose is. Purpose gives you direction to where you should go. Ever since I have this  fire to inspire the next person. I hope this book inspires someone to follow their dreams.